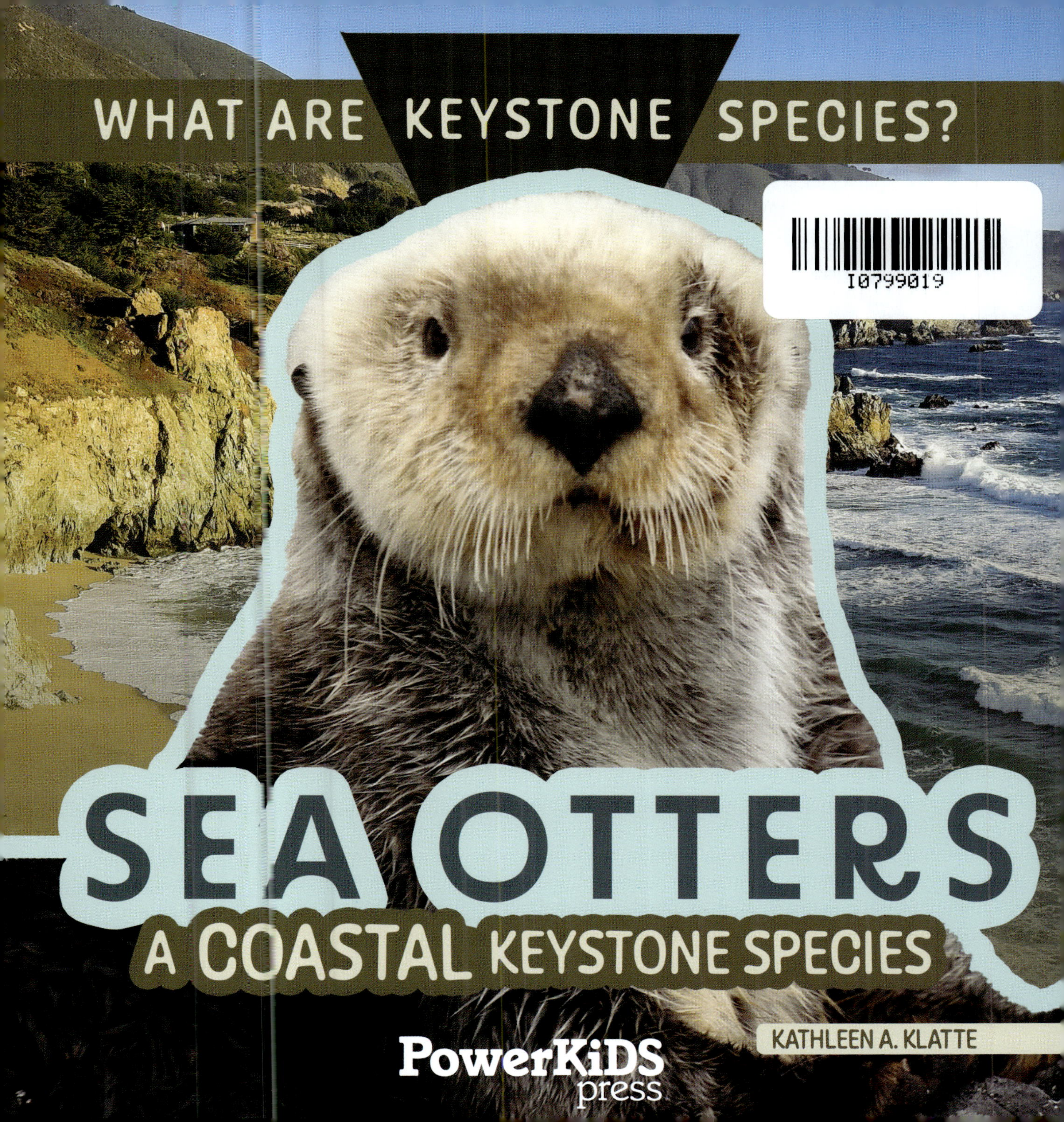

WHAT ARE KEYSTONE SPECIES?

SEA OTTERS

A COASTAL KEYSTONE SPECIES

KATHLEEN A. KLATTE

PowerKiDS press

Published in 2025 by The Rosen Publishing Group, Inc.
2544 Clinton Street, Buffalo, NY 14224

First Edition

Editor: Theresa Emminizer
Book Design: Tanya Dellaccio Keeney

Photo Credits: Cover tryton2011/Shutterstock.com; p. 5 Snorre Roberg/Shutterstock.com; p. 7 [illegible]elpixel/Shutterstock.com; p. 9 Manuel Balesteri/Shutterstock.com; p. 11 Ethan Daniels/Shutterstock.com; p. 13 David A Litman/Shutterstock.com; p. 15 Andrew b Stowe/Shutterstock.com; p. 17 Rich Carey/Shutterstock.com; p. 19 Wildnerdpix/Shutterstock.com; p. 21 EyesTravelling/Shutterstock.com.

Cataloging-in-Publication Data

Names: Klatte, Kathleen A.
Title: Sea otters: a coastal keystone species / Kathleen A. Klatte.
Description: Buffalo, NY : PowerKids Press, 2025. | Series: What are keystone species? | Includes glossary and index.
Identifiers: ISBN 9781499446623 (pbk.) | ISBN 9781499446630 (library bound) | ISBN 9781499446647 (ebook)
Subjects: LCSH: Sea otter–Juvenile literature. | Sea otter–Conservation–Juvenile literature. | Keystone species–Juvenile literature. | Marine ecology–Juvenile literature.
Classification: LCC QL737.C25 K525 2025 | DDC 599.769'5–dc23

Manufactured in the United States of America

CPSIA Compliance Information: Batch #CSPK25. For further information contact Rosen Publishing at 1-800-237-9932.

CONTENTS

SPECIAL SEA OTTERS

Sea otters are a keystone species. This means that they're very important to the place where they live. Sea otters are predators. They protect their habitat, or natural home, by keeping the number of prey animals that live there in check.

Sea otters live in underwater **ecosystems** called kelp forests. Kelp is a kind of seaweed. Without sea otters, urchins would eat all the kelp. None would be left for other animals to eat or make their homes in.

IN FACT!

SEA URCHINS ARE SPINY CREATURES THAT LIVE ON THE OCEAN FLOOR. THEY EAT PLANTS, LIKE KELP. A FEMALE SEA URCHIN CAN LAY MILLIONS OF EGGS AT A TIME.

WHEN SEA URCHINS EAT ALL THE KELP, ALL THE OTHER KINDS OF ANIMALS GO AWAY. THIS FORMS AN AREA CALLED AN URCHIN BARREN.

MAMMALS

Otters are mammals. Mammals are warm-blooded and give birth to live babies. They also make milk to feed babies. Sea otters have one baby at a time. A baby otter is usually called a pup. Sometimes they're called kittens or kits.

Mammals also usually have hair or fur. Otters have thick brown fur that keeps them warm in the cold water. Sea otters have the thickest fur of any mammal. Otter's special fur can trap air and help otters float!

IN FACT!

OTTERS ARE CARNIVORES. THIS MEANS THEY EAT MEAT, NOT PLANTS.

WARM-BLOODED ANIMALS KEEP THE SAME BODY TEMPERATURE, EVEN IF THEY'RE SWIMMING IN COLD WATER.

SEA OTTER FAMILIES

Sea otter pups are about 22 to 24 inches (56 to 61 cm) long. They weigh about 3 to 5 pounds (1.4 to 2.3 kg) at birth. Sea otter pups are born with their eyes open and their first teeth grown in. Pups drink milk for six to eight months, even though they start to eat solid food at about four weeks.

They're also born with a full coat of fur that helps them float. They learn to swim at about four weeks of age and dive at six weeks.

IN FACT!

A GROUP OF OTTERS FLOATING TOGETHER IS CALLED A RAFT. RAFTS TEND TO BE MADE UP OF EITHER ALL GIRLS OR ALL BOYS.

MOTHER OTTERS CARRY THEIR PUPS ON THEIR BELLY SO THEY CAN NURSE. MOTHERS SPEND A LOT OF TIME **GROOMING** THEIR BABIES

WHAT'S A KELP FOREST?

Kelp is a kind of brown algae. Algae are plant-like living things with no leaves or stems that grow near water. Kelp grows in coastal waters. It can grow to 200 feet (61 m)!

Kelp makes its own food through a process, or set of steps, called photosynthesis (foh-toh-SIHN-thuh-suhs). Through photosynthesis, kelp turns water and **carbon dioxide** into food. This is the same process plants use. Like plants, kelp needs sunlight. This is why it grows in water that's shallow, or not very deep.

IN FACT!

MARINE, OR OCEAN, ALGAE, INCLUDING KELP, MAKE 70 TO 80 PERCENT OF THE **OXYGEN** WE BREATH.

KELP GROW TOGETHER IN GROUPS. THEY LOOK A LOT LIKE TREES GROWING IN A FOREST ON LAND.

WHAT ELSE LIVES THERE?

A kelp forest is home to many different species, or kinds, of animals. Many fish lay eggs and raise their babies there. Kelp blades—which are sort of like leaves—make good hiding places for baby fish. Sea birds catch fish and crabs in the kelp forest.

Many kinds of **invertebrates** live there too. Sponges, crabs, and snails live on the ocean floor. So do sea stars and urchins. Like otters, sea stars help control the urchin **population**.

IN FACT!

MANY ANIMALS IN THE KELP FOREST EAT PLANKTON. THESE ARE TINY PLANTS AND ANIMALS THAT FLOAT IN THE OCEAN.

FISH LAY THEIR EGGS ON SEA ALGAE LIKE KELP TO KEEP THE EGGS SAFE FROM BEING EATEN OR FLOATING AWAY.

PREDATOR AND PREY

There are many predators that hunt in the kelp forest. Some kinds of animals are both predators and prey. Sharks eat fish, fish eggs, and crabs. Giant sea bass eat all sorts of fish and marine life, including small sharks!

Octopuses eat snails and crabs. Water birds dive underwater to catch crabs, worms, and small fish. The California sheephead is a fish with strong teeth to bite through the shells of urchins, crabs, and lobsters.

IN FACT!

ANIMALS THAT EAT OTHER ANIMALS AREN'T BEING MEAN. THEY'RE HELPING THE ECOSYSTEM! WITHOUT PREDATORS, PREY ANIMALS WOULD EAT UP ALL THE PLANTS. THEN THEY WOULDN'T HAVE ANY MORE FOOD.

A HEALTHY KELP FOREST IS FILLED WITH MANY DIFFERENT ANIMALS.

SEA OTTERS IN DANGER

People are the biggest threat, or danger, to sea otters. People used to hunt otters for their fur. Today there are laws protecting otters and other marine mammals. However, sea otters are considered endangered. An endangered species is one that might become extinct.

Sea otters can be hurt or killed by boats. They can also become tangled in fishing nets and drown. Oil spills are especially bad. Oil damages otters' fur so it doesn't keep them warm and dry. It can also **poison** their food.

GARBAGE—ESPECIALLY PLASTIC—IN THE WATER IS BAD FOR EVERYTHING THAT LIVES THERE.

IN FACT!

WHEN A SPECIES IS EXTINCT, IT MEANS THERE ARE NO MORE LIVING MEMBERS OF THE SPECIES. A THREATENED SPECIES IS ONE THAT MIGHT BECOME ENDANGERED.

WHAT IF THERE WERE NO SEA OTTERS?

If there were no otters, purple sea urchins would eat all the kelp. If the kelp were gone, all the other animals would go away. Eventually, the entire ocean floor would be covered with urchins. The urchins would gobble up any new kelp that started to grow.

Urchins can live for many years, even if there's no kelp for them to eat. Although some people like to eat urchins, ones that are **starving** don't taste good, so fishermen can't sell them to restaurants.

OTTERS AREN'T JUST CUTE. THEY'RE VERY HELPFUL ANIMALS!

IN FACT!

SEA STARS ARE ALSO KEYSTONE PREDATORS OF SEA URCHINS. HOWEVER, A TERRIBLE DISEASE, OR SICKNESS, HAS KILLED MANY OF THEM IN THE LAST FEW YEARS.

HOW CAN WE HELP?

There are many things people can do to help otters. The best thing we can do is to not **pollute** the water. Everything that lives on Earth, including people, needs clean water.

We can also learn to be careful of otters. Otters are wild animals. When people go where otters live in the wild, they need to leave them alone. It's OK to take pictures or watch them from far away. It's not OK to try to touch them or play with them.

IN FACT!

OTTERS USE ROCKS TO OPEN THE SHELLS OF THEIR DINNER.

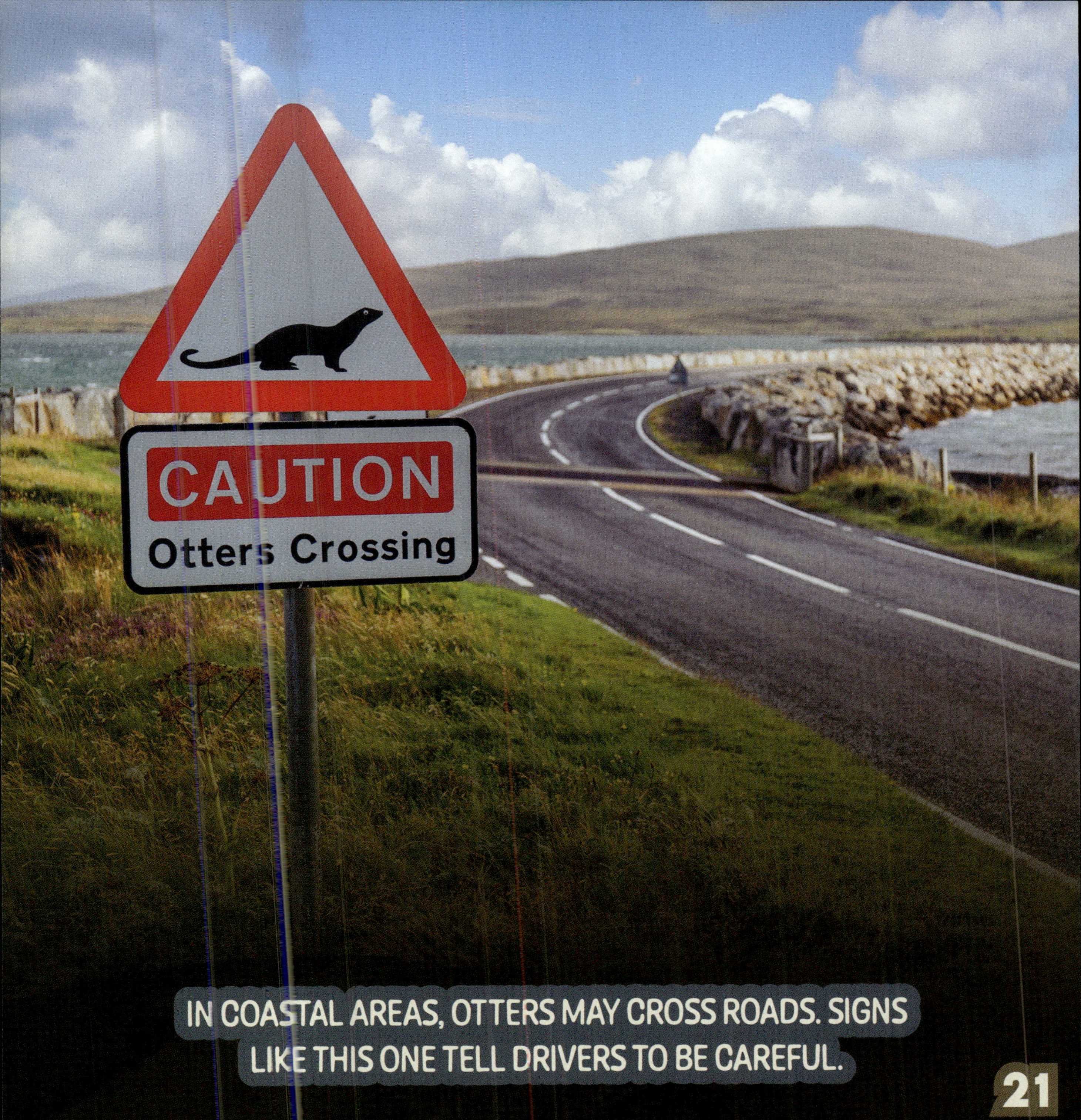

IN COASTAL AREAS, OTTERS MAY CROSS ROADS. SIGNS LIKE THIS ONE TELL DRIVERS TO BE CAREFUL.

GLOSSARY

carbon dioxide: A gas breathed out by animals.

ecosystem: A natural community of living and nonliving things.

grooming: The process of cleaning the fur, skin, or feathers.

invertebrate: A type of animal that does not have a backbone.

oxygen: A colorless, odorless gas that many animals, including people, need to breathe.

poison: A harmful kind of matter that can kill a living thing.

pollute: To make (land, water, air, etc.) dirty and not safe or suitable to use.

population: The number of animals in a species that live in a place.

starve: To suffer or die from lack of food: to suffer extreme hunger.

FOR MORE INFORMATION

BOOKS

Lock, Deborah. *Sea Otters*. New York, NY: DK Publishing, 2023.

Pringle, Laurence. *The Secret Life of the Sea Otter*. New York, NY: Astra Young Readers, 2022.

WEBSITES

Kelp
www.sailorsforthesea.org/programs/kelp
Kids Environmental Lesson Plans (KELP) is Oceana's page of lessons and activities for kids of various ages.

Young Ocean Explorers
www.youngoceanexplorers.com/yoe/quiz/234277314721#cplayer
Young Ocean Explorers has a page of fun kids' activities about ocean conservation.

Publisher's note to educators and parents: Our editors have carefully reviewed these websites to ensure that they are suitable for students. Many websites change frequently, however, and we cannot guarantee that a site's future contents will continue to meet our high standards of quality and educational value. Be advised that students should be closely supervised whenever they access the internet.

INDEX